Clients Name:	
Clients Phone No. Email, skype etc.:	
Date of meeting:	
Brief Description:	
Project Name/ REF:	
Area of Property:	Action Points/ Jobs/ Outcomes/ Tasks/ etc:

AF808164

Additional Notes/ Initial Thoughts/ Potential Problems/ Follow Ups etc:

Clients Name:	
Clients Phone No. Email, skype etc.:	
Date of meeting:	
Brief Description:	
Project Name/ REF:	
Area of Property:	Action Points/ Jobs/ Outcomes/ Tasks/ etc:

Additional Notes/ Initial Thoughts/ Potential Problems/ Follow Ups etc:

Clients Name:	
Clients Phone No. Email, skype etc.:	
Date of meeting:	
Brief Description:	
Project Name/ REF:	
Area of Property:	Action Points/ Jobs/ Outcomes/ Tasks/ etc:

Additional Notes/ Initial Thoughts/ Potential Problems/ Follow Ups etc:

Clients Name:	
Clients Phone No. Email, skype etc.:	
Date of meeting:	
Brief Description:	
Project Name/ REF:	
Area of Property:	Action Points/ Jobs/ Outcomes/ Tasks/ etc:

Additional Notes/ Initial Thoughts/ Potential Problems/ Follow Ups etc:

Clients Name:	
Clients Phone No. Email, skype etc.:	
Date of meeting:	
Brief Description:	
Project Name/ REF:	
Area of Property:	Action Points/ Jobs/ Outcomes/ Tasks/ etc:

Additional Notes/ Initial Thoughts/ Potential Problems/ Follow Ups etc:

Clients Name:	
Clients Phone No. Email, skype etc.:	
Date of meeting:	
Brief Description:	
Project Name/ REF:	
Area of Property:	Action Points/ Jobs/ Outcomes/ Tasks/ etc:

Additional Notes/ Initial Thoughts/ Potential Problems/ Follow Ups etc:

Clients Name:	
Clients Phone No. Email, skype etc.:	
Date of meeting:	
Brief Description:	
Project Name/ REF:	
Area of Property:	Action Points/ Jobs/ Outcomes/ Tasks/ etc:

Additional Notes/ Initial Thoughts/ Potential Problems/ Follow Ups etc:

Clients Name:	
Clients Phone No. Email, skype etc.:	
Date of meeting:	
Brief Description:	
Project Name/ REF:	
Area of Property:	Action Points/ Jobs/ Outcomes/ Tasks/ etc:

Additional Notes/ Initial Thoughts/ Potential Problems/ Follow Ups etc:

Clients Name:	
Clients Phone No. Email, skype etc.:	
Date of meeting:	
Brief Description:	
Project Name/ REF:	
Area of Property:	Action Points/ Jobs/ Outcomes/ Tasks/ etc:

Additional Notes/ Initial Thoughts/ Potential Problems/ Follow Ups etc:

Clients Name:	
Clients Phone No. Email, skype etc.:	
Date of meeting:	
Brief Description:	
Project Name/ REF:	
Area of Property:	Action Points/ Jobs/ Outcomes/ Tasks/ etc:

Additional Notes/ Initial Thoughts/ Potential Problems/ Follow Ups etc:

Clients Name:	
Clients Phone No. Email, skype etc.:	
Date of meeting:	
Brief Description:	
Project Name/ REF:	
Area of Property:	Action Points/ Jobs/ Outcomes/ Tasks/ etc:

Additional Notes/ Initial Thoughts/ Potential Problems/ Follow Ups etc:

Clients Name:	
Clients Phone No. Email, skype etc.:	
Date of meeting:	
Brief Description:	
Project Name/ REF:	
Area of Property:	Action Points/ Jobs/ Outcomes/ Tasks/ etc:

Additional Notes/ Initial Thoughts/ Potential Problems/ Follow Ups etc:

Clients Name:	
Clients Phone No. Email, skype etc.:	
Date of meeting:	
Brief Description:	
Project Name/ REF:	
Area of Property:	Action Points/ Jobs/ Outcomes/ Tasks/ etc:

Additional Notes/ Initial Thoughts/ Potential Problems/ Follow Ups etc:

Clients Name:	
Clients Phone No. Email, skype etc.:	
Date of meeting:	
Brief Description:	
Project Name/ REF:	
Area of Property:	Action Points/ Jobs/ Outcomes/ Tasks/ etc:

Additional Notes/ Initial Thoughts/ Potential Problems/ Follow Ups etc:

Clients Name:	
Clients Phone No. Email, skype etc.:	
Date of meeting:	
Brief Description:	
Project Name/ REF:	
Area of Property:	Action Points/ Jobs/ Outcomes/ Tasks/ etc:

Additional Notes/ Initial Thoughts/ Potential Problems/ Follow Ups etc:

Clients Name:	
Clients Phone No. Email, skype etc.:	
Date of meeting:	
Brief Description:	
Project Name/ REF:	
Area of Property:	Action Points/ Jobs/ Outcomes/ Tasks/ etc:

Additional Notes/ Initial Thoughts/ Potential Problems/ Follow Ups etc:

Clients Name:	
Clients Phone No. Email, skype etc.:	
Date of meeting:	
Brief Description:	
Project Name/ REF:	
Area of Property:	Action Points/ Jobs/ Outcomes/ Tasks/ etc:

Additional Notes/ Initial Thoughts/ Potential Problems/ Follow Ups etc:

Clients Name:	
Clients Phone No. Email, skype etc.:	
Date of meeting:	
Brief Description:	
Project Name/ REF:	
Area of Property:	Action Points/ Jobs/ Outcomes/ Tasks/ etc:

Additional Notes/ Initial Thoughts/ Potential Problems/ Follow Ups etc:

Clients Name:	
Clients Phone No. Email, skype etc.:	
Date of meeting:	
Brief Description:	
Project Name/ REF:	
Area of Property:	Action Points/ Jobs/ Outcomes/ Tasks/ etc:

Additional Notes/ Initial Thoughts/ Potential Problems/ Follow Ups etc:

Clients Name:	
Clients Phone No. Email, skype etc.:	
Date of meeting:	
Brief Description:	
Project Name/ REF:	
Area of Property:	Action Points/ Jobs/ Outcomes/ Tasks/ etc:

Additional Notes/ Initial Thoughts/ Potential Problems/ Follow Ups etc:

Clients Name:	
Clients Phone No. Email, skype etc.:	
Date of meeting:	
Brief Description:	
Project Name/ REF:	
Area of Property:	Action Points/ Jobs/ Outcomes/ Tasks/ etc:

Additional Notes/ Initial Thoughts/ Potential Problems/ Follow Ups etc:

Clients Name:	
Clients Phone No. Email, skype etc.:	
Date of meeting:	
Brief Description:	
Project Name/ REF:	
Area of Property:	Action Points/ Jobs/ Outcomes/ Tasks/ etc:

Additional Notes/ Initial Thoughts/ Potential Problems/ Follow Ups etc:

Clients Name:	
Clients Phone No. Email, skype etc.:	
Date of meeting:	
Brief Description:	
Project Name/ REF:	
Area of Property:	Action Points/ Jobs/ Outcomes/ Tasks/ etc:

Additional Notes/ Initial Thoughts/ Potential Problems/ Follow Ups etc:

Clients Name:	
Clients Phone No. Email, skype etc.:	
Date of meeting:	
Brief Description:	
Project Name/ REF:	
Area of Property:	Action Points/ Jobs/ Outcomes/ Tasks/ etc:

Additional Notes/ Initial Thoughts/ Potential Problems/ Follow Ups etc:

Clients Name:	
Clients Phone No. Email, skype etc.:	
Date of meeting:	
Brief Description:	
Project Name/ REF:	
Area of Property:	Action Points/ Jobs/ Outcomes/ Tasks/ etc:

Additional Notes/ Initial Thoughts/ Potential Problems/ Follow Ups etc:

Clients Name:	
Clients Phone No. Email, skype etc.:	
Date of meeting:	
Brief Description:	
Project Name/ REF:	
Area of Property:	Action Points/ Jobs/ Outcomes/ Tasks/ etc:

Additional Notes/ Initial Thoughts/ Potential Problems/ Follow Ups etc:

Clients Name:	
Clients Phone No. Email, skype etc.:	
Date of meeting:	
Brief Description:	
Project Name/ REF:	
Area of Property:	Action Points/ Jobs/ Outcomes/ Tasks/ etc:

Additional Notes/ Initial Thoughts/ Potential Problems/ Follow Ups etc:

Clients Name:	
Clients Phone No. Email, skype etc.:	
Date of meeting:	
Brief Description:	
Project Name/ REF:	
Area of Property:	Action Points/ Jobs/ Outcomes/ Tasks/ etc:
Additional Notes/ Initial Thoughts/ Potential Problems/ Follow Ups etc:	

Clients Name:	
Clients Phone No. Email, skype etc.:	
Date of meeting:	
Brief Description:	
Project Name/ REF:	
Area of Property:	Action Points/ Jobs/ Outcomes/ Tasks/ etc:

Additional Notes/ Initial Thoughts/ Potential Problems/ Follow Ups etc:

Clients Name:	
Clients Phone No. Email, skype etc.:	
Date of meeting:	
Brief Description:	
Project Name/ REF:	
Area of Property:	Action Points/ Jobs/ Outcomes/ Tasks/ etc:

Additional Notes/ Initial Thoughts/ Potential Problems/ Follow Ups etc:

Clients Name:	
Clients Phone No. Email, skype etc.:	
Date of meeting:	
Brief Description:	
Project Name/ REF:	
Area of Property:	Action Points/ Jobs/ Outcomes/ Tasks/ etc:

Additional Notes/ Initial Thoughts/ Potential Problems/ Follow Ups etc:

Clients Name:	
Clients Phone No. Email, skype etc.:	
Date of meeting:	
Brief Description:	
Project Name/ REF:	
Area of Property:	Action Points/ Jobs/ Outcomes/ Tasks/ etc:

Additional Notes/ Initial Thoughts/ Potential Problems/ Follow Ups etc:

Clients Name:	
Clients Phone No. Email, skype etc.:	
Date of meeting:	
Brief Description:	
Project Name/ REF:	
Area of Property:	Action Points/ Jobs/ Outcomes/ Tasks/ etc:

Additional Notes/ Initial Thoughts/ Potential Problems/ Follow Ups etc:

Clients Name:	
Clients Phone No. Email, skype etc.:	
Date of meeting:	
Brief Description:	
Project Name/ REF:	
Area of Property:	Action Points/ Jobs/ Outcomes/ Tasks/ etc:

Additional Notes/ Initial Thoughts/ Potential Problems/ Follow Ups etc:

Clients Name:	
Clients Phone No. Email, skype etc.:	
Date of meeting:	
Brief Description:	
Project Name/ REF:	
Area of Property:	Action Points/ Jobs/ Outcomes/ Tasks/ etc:

Additional Notes/ Initial Thoughts/ Potential Problems/ Follow Ups etc:

Clients Name:	
Clients Phone No. Email, skype etc.:	
Date of meeting:	
Brief Description:	
Project Name/ REF:	
Area of Property:	Action Points/ Jobs/ Outcomes/ Tasks/ etc:

Additional Notes/ Initial Thoughts/ Potential Problems/ Follow Ups etc:

Clients Name:	
Clients Phone No. Email, skype etc.:	
Date of meeting:	
Brief Description:	
Project Name/ REF:	
Area of Property:	Action Points/ Jobs/ Outcomes/ Tasks/ etc:

Additional Notes/ Initial Thoughts/ Potential Problems/ Follow Ups etc:

Clients Name:	
Clients Phone No. Email, skype etc.:	
Date of meeting:	
Brief Description:	
Project Name/ REF:	
Area of Property:	Action Points/ Jobs/ Outcomes/ Tasks/ etc:

Additional Notes/ Initial Thoughts/ Potential Problems/ Follow Ups etc:

Clients Name:	
Clients Phone No. Email, skype etc.:	
Date of meeting:	
Brief Description:	
Project Name/ REF:	
Area of Property:	Action Points/ Jobs/ Outcomes/ Tasks/ etc:

Additional Notes/ Initial Thoughts/ Potential Problems/ Follow Ups etc:

Clients Name:	
Clients Phone No. Email, skype etc.:	
Date of meeting:	
Brief Description:	
Project Name/ REF:	
Area of Property:	Action Points/ Jobs/ Outcomes/ Tasks/ etc:

Additional Notes/ Initial Thoughts/ Potential Problems/ Follow Ups etc:

Clients Name:	
Clients Phone No. Email, skype etc.:	
Date of meeting:	
Brief Description:	
Project Name/ REF:	
Area of Property:	Action Points/ Jobs/ Outcomes/ Tasks/ etc:

Additional Notes/ Initial Thoughts/ Potential Problems/ Follow Ups etc:

Clients Name:	
Clients Phone No. Email, skype etc.:	
Date of meeting:	
Brief Description:	
Project Name/ REF:	
Area of Property:	Action Points/ Jobs/ Outcomes/ Tasks/ etc:

Additional Notes/ Initial Thoughts/ Potential Problems/ Follow Ups etc:

Clients Name:	
Clients Phone No. Email, skype etc.:	
Date of meeting:	
Brief Description:	
Project Name/ REF:	
Area of Property:	Action Points/ Jobs/ Outcomes/ Tasks/ etc:

Additional Notes/ Initial Thoughts/ Potential Problems/ Follow Ups etc:

Clients Name:	
Clients Phone No. Email, skype etc.:	
Date of meeting:	
Brief Description:	
Project Name/ REF:	
Area of Property:	Action Points/ Jobs/ Outcomes/ Tasks/ etc:

Additional Notes/ Initial Thoughts/ Potential Problems/ Follow Ups etc:

Clients Name:	
Clients Phone No. Email, skype etc.:	
Date of meeting:	
Brief Description:	
Project Name/ REF:	
Area of Property:	Action Points/ Jobs/ Outcomes/ Tasks/ etc:

Additional Notes/ Initial Thoughts/ Potential Problems/ Follow Ups etc:

Clients Name:	
Clients Phone No. Email, skype etc.:	
Date of meeting:	
Brief Description:	
Project Name/ REF:	
Area of Property:	Action Points/ Jobs/ Outcomes/ Tasks/ etc:

Additional Notes/ Initial Thoughts/ Potential Problems/ Follow Ups etc:

Clients Name:	
Clients Phone No. Email, skype etc.:	
Date of meeting:	
Brief Description:	
Project Name/ REF:	
Area of Property:	Action Points/ Jobs/ Outcomes/ Tasks/ etc:

Additional Notes/ Initial Thoughts/ Potential Problems/ Follow Ups etc:

Clients Name:	
Clients Phone No. Email, skype etc.:	
Date of meeting:	
Brief Description:	
Project Name/ REF:	
Area of Property:	Action Points/ Jobs/ Outcomes/ Tasks/ etc:

Additional Notes/ Initial Thoughts/ Potential Problems/ Follow Ups etc:

Clients Name:	
Clients Phone No. Email, skype etc.:	
Date of meeting:	
Brief Description:	
Project Name/ REF:	
Area of Property:	Action Points/ Jobs/ Outcomes/ Tasks/ etc:

Additional Notes/ Initial Thoughts/ Potential Problems/ Follow Ups etc:

Clients Name:	
Clients Phone No. Email, skype etc.:	
Date of meeting:	
Brief Description:	
Project Name/ REF:	
Area of Property:	Action Points/ Jobs/ Outcomes/ Tasks/ etc:

Additional Notes/ Initial Thoughts/ Potential Problems/ Follow Ups etc:

Clients Name:	
Clients Phone No. Email, skype etc.:	
Date of meeting:	
Brief Description:	
Project Name/ REF:	
Area of Property:	Action Points/ Jobs/ Outcomes/ Tasks/ etc:

Additional Notes/ Initial Thoughts/ Potential Problems/ Follow Ups etc:

Clients Name:	
Clients Phone No. Email, skype etc.:	
Date of meeting:	
Brief Description:	
Project Name/ REF:	
Area of Property:	Action Points/ Jobs/ Outcomes/ Tasks/ etc:

Additional Notes/ Initial Thoughts/ Potential Problems/ Follow Ups etc:

Clients Name:	
Clients Phone No. Email, skype etc.:	
Date of meeting:	
Brief Description:	
Project Name/ REF:	
Area of Property:	Action Points/ Jobs/ Outcomes/ Tasks/ etc:

Additional Notes/ Initial Thoughts/ Potential Problems/ Follow Ups etc:

Clients Name:	
Clients Phone No. Email, skype etc.:	
Date of meeting:	
Brief Description:	
Project Name/ REF:	
Area of Property:	Action Points/ Jobs/ Outcomes/ Tasks/ etc:

Additional Notes/ Initial Thoughts/ Potential Problems/ Follow Ups etc:

Clients Name:	
Clients Phone No. Email, skype etc.:	
Date of meeting:	
Brief Description:	
Project Name/ REF:	
Area of Property:	Action Points/ Jobs/ Outcomes/ Tasks/ etc:

Additional Notes/ Initial Thoughts/ Potential Problems/ Follow Ups etc:

Clients Name:	
Clients Phone No. Email, skype etc.:	
Date of meeting:	
Brief Description:	
Project Name/ REF:	
Area of Property:	Action Points/ Jobs/ Outcomes/ Tasks/ etc:

Additional Notes/ Initial Thoughts/ Potential Problems/ Follow Ups etc:

Clients Name:	
Clients Phone No. Email, skype etc.:	
Date of meeting:	
Brief Description:	
Project Name/ REF:	
Area of Property:	Action Points/ Jobs/ Outcomes/ Tasks/ etc:

Additional Notes/ Initial Thoughts/ Potential Problems/ Follow Ups etc:

Clients Name:	
Clients Phone No. Email, skype etc.:	
Date of meeting:	
Brief Description:	
Project Name/ REF:	
Area of Property:	Action Points/ Jobs/ Outcomes/ Tasks/ etc:

Additional Notes/ Initial Thoughts/ Potential Problems/ Follow Ups etc:

Clients Name:	
Clients Phone No. Email, skype etc.:	
Date of meeting:	
Brief Description:	
Project Name/ REF:	
Area of Property:	Action Points/ Jobs/ Outcomes/ Tasks/ etc:

Additional Notes/ Initial Thoughts/ Potential Problems/ Follow Ups etc:

Clients Name:	
Clients Phone No. Email, skype etc.:	
Date of meeting:	
Brief Description:	
Project Name/ REF:	
Area of Property:	Action Points/ Jobs/ Outcomes/ Tasks/ etc:

Additional Notes/ Initial Thoughts/ Potential Problems/ Follow Ups etc:

Clients Name:	
Clients Phone No. Email, skype etc.:	
Date of meeting:	
Brief Description:	
Project Name/ REF:	
Area of Property:	Action Points/ Jobs/ Outcomes/ Tasks/ etc:

Additional Notes/ Initial Thoughts/ Potential Problems/ Follow Ups etc:

Clients Name:	
Clients Phone No. Email, skype etc.:	
Date of meeting:	
Brief Description:	
Project Name/ REF:	
Area of Property:	Action Points/ Jobs/ Outcomes/ Tasks/ etc:

Additional Notes/ Initial Thoughts/ Potential Problems/ Follow Ups etc:

Clients Name:	
Clients Phone No. Email, skype etc.:	
Date of meeting:	
Brief Description:	
Project Name/ REF:	
Area of Property:	Action Points/ Jobs/ Outcomes/ Tasks/ etc:

Additional Notes/ Initial Thoughts/ Potential Problems/ Follow Ups etc:

Clients Name:	
Clients Phone No. Email, skype etc.:	
Date of meeting:	
Brief Description:	
Project Name/ REF:	
Area of Property:	Action Points/ Jobs/ Outcomes/ Tasks/ etc:

Additional Notes/ Initial Thoughts/ Potential Problems/ Follow Ups etc:

Clients Name:	
Clients Phone No. Email, skype etc.:	
Date of meeting:	
Brief Description:	
Project Name/ REF:	
Area of Property:	Action Points/ Jobs/ Outcomes/ Tasks/ etc:

Additional Notes/ Initial Thoughts/ Potential Problems/ Follow Ups etc:

Clients Name:	
Clients Phone No. Email, skype etc.:	
Date of meeting:	
Brief Description:	
Project Name/ REF:	
Area of Property:	Action Points/ Jobs/ Outcomes/ Tasks/ etc:

Additional Notes/ Initial Thoughts/ Potential Problems/ Follow Ups etc:

Clients Name:	
Clients Phone No. Email, skype etc.:	
Date of meeting:	
Brief Description:	
Project Name/ REF:	
Area of Property:	Action Points/ Jobs/ Outcomes/ Tasks/ etc:

Additional Notes/ Initial Thoughts/ Potential Problems/ Follow Ups etc:

Clients Name:	
Clients Phone No. Email, skype etc.:	
Date of meeting:	
Brief Description:	
Project Name/ REF:	
Area of Property:	Action Points/ Jobs/ Outcomes/ Tasks/ etc:

Additional Notes/ Initial Thoughts/ Potential Problems/ Follow Ups etc:

Clients Name:	
Clients Phone No. Email, skype etc.:	
Date of meeting:	
Brief Description:	
Project Name/ REF:	
Area of Property:	Action Points/ Jobs/ Outcomes/ Tasks/ etc:

Additional Notes/ Initial Thoughts/ Potential Problems/ Follow Ups etc:

Clients Name:	
Clients Phone No. Email, skype etc.:	
Date of meeting:	
Brief Description:	
Project Name/ REF:	
Area of Property:	Action Points/ Jobs/ Outcomes/ Tasks/ etc:

Additional Notes/ Initial Thoughts/ Potential Problems/ Follow Ups etc:

Clients Name:	
Clients Phone No. Email, skype etc.:	
Date of meeting:	
Brief Description:	
Project Name/ REF:	
Area of Property:	Action Points/ Jobs/ Outcomes/ Tasks/ etc:

Additional Notes/ Initial Thoughts/ Potential Problems/ Follow Ups etc:

Clients Name:	
Clients Phone No. Email, skype etc.:	
Date of meeting:	
Brief Description:	
Project Name/ REF:	
Area of Property:	Action Points/ Jobs/ Outcomes/ Tasks/ etc:

Additional Notes/ Initial Thoughts/ Potential Problems/ Follow Ups etc:

Clients Name:	
Clients Phone No. Email, skype etc.:	
Date of meeting:	
Brief Description:	
Project Name/ REF:	
Area of Property:	Action Points/ Jobs/ Outcomes/ Tasks/ etc:

Additional Notes/ Initial Thoughts/ Potential Problems/ Follow Ups etc:

Clients Name:	
Clients Phone No. Email, skype etc.:	
Date of meeting:	
Brief Description:	
Project Name/ REF:	
Area of Property:	Action Points/ Jobs/ Outcomes/ Tasks/ etc:

Additional Notes/ Initial Thoughts/ Potential Problems/ Follow Ups etc:

Clients Name:	
Clients Phone No. Email, skype etc.:	
Date of meeting:	
Brief Description:	
Project Name/ REF:	
Area of Property:	Action Points/ Jobs/ Outcomes/ Tasks/ etc:

Additional Notes/ Initial Thoughts/ Potential Problems/ Follow Ups etc:

Clients Name:	
Clients Phone No. Email, skype etc.:	
Date of meeting:	
Brief Description:	
Project Name/ REF:	
Area of Property:	Action Points/ Jobs/ Outcomes/ Tasks/ etc:

Additional Notes/ Initial Thoughts/ Potential Problems/ Follow Ups etc:

Clients Name:	
Clients Phone No. Email, skype etc.:	
Date of meeting:	
Brief Description:	
Project Name/ REF:	
Area of Property:	Action Points/ Jobs/ Outcomes/ Tasks/ etc:

Additional Notes/ Initial Thoughts/ Potential Problems/ Follow Ups etc:

Clients Name:	
Clients Phone No. Email, skype etc.:	
Date of meeting:	
Brief Description:	
Project Name/ REF:	
Area of Property:	Action Points/ Jobs/ Outcomes/ Tasks/ etc:

Additional Notes/ Initial Thoughts/ Potential Problems/ Follow Ups etc:

Clients Name:	
Clients Phone No. Email, skype etc.:	
Date of meeting:	
Brief Description:	
Project Name/ REF:	
Area of Property:	Action Points/ Jobs/ Outcomes/ Tasks/ etc:

Additional Notes/ Initial Thoughts/ Potential Problems/ Follow Ups etc:

Clients Name:	
Clients Phone No. Email, skype etc.:	
Date of meeting:	
Brief Description:	
Project Name/ REF:	
Area of Property:	Action Points/ Jobs/ Outcomes/ Tasks/ etc:

Additional Notes/ Initial Thoughts/ Potential Problems/ Follow Ups etc:

Clients Name:	
Clients Phone No. Email, skype etc.:	
Date of meeting:	
Brief Description:	
Project Name/ REF:	
Area of Property:	Action Points/ Jobs/ Outcomes/ Tasks/ etc:

Additional Notes/ Initial Thoughts/ Potential Problems/ Follow Ups etc:

Clients Name:	
Clients Phone No. Email, skype etc.:	
Date of meeting:	
Brief Description:	
Project Name/ REF:	
Area of Property:	Action Points/ Jobs/ Outcomes/ Tasks/ etc:

Additional Notes/ Initial Thoughts/ Potential Problems/ Follow Ups etc:

Clients Name:	
Clients Phone No. Email, skype etc.:	
Date of meeting:	
Brief Description:	
Project Name/ REF:	
Area of Property:	Action Points/ Jobs/ Outcomes/ Tasks/ etc:

Additional Notes/ Initial Thoughts/ Potential Problems/ Follow Ups etc:

Clients Name:	
Clients Phone No. Email, skype etc.:	
Date of meeting:	
Brief Description:	
Project Name/ REF:	
Area of Property:	Action Points/ Jobs/ Outcomes/ Tasks/ etc:

Additional Notes/ Initial Thoughts/ Potential Problems/ Follow Ups etc:

Clients Name:	
Clients Phone No. Email, skype etc.:	
Date of meeting:	
Brief Description:	
Project Name/ REF:	
Area of Property:	Action Points/ Jobs/ Outcomes/ Tasks/ etc:

Additional Notes/ Initial Thoughts/ Potential Problems/ Follow Ups etc:

Clients Name:	
Clients Phone No. Email, skype etc.:	
Date of meeting:	
Brief Description:	
Project Name/ REF:	
Area of Property:	Action Points/ Jobs/ Outcomes/ Tasks/ etc:

Additional Notes/ Initial Thoughts/ Potential Problems/ Follow Ups etc:

Clients Name:	
Clients Phone No. Email, skype etc.:	
Date of meeting:	
Brief Description:	
Project Name/ REF:	
Area of Property:	Action Points/ Jobs/ Outcomes/ Tasks/ etc:

Additional Notes/ Initial Thoughts/ Potential Problems/ Follow Ups etc:

Clients Name:	
Clients Phone No. Email, skype etc.:	
Date of meeting:	
Brief Description:	
Project Name/ REF:	
Area of Property:	Action Points/ Jobs/ Outcomes/ Tasks/ etc:

Additional Notes/ Initial Thoughts/ Potential Problems/ Follow Ups etc:

Clients Name:	
Clients Phone No. Email, skype etc.:	
Date of meeting:	
Brief Description:	
Project Name/ REF:	
Area of Property:	Action Points/ Jobs/ Outcomes/ Tasks/ etc:

Additional Notes/ Initial Thoughts/ Potential Problems/ Follow Ups etc:

Clients Name:	
Clients Phone No. Email, skype etc.:	
Date of meeting:	
Brief Description:	
Project Name/ REF:	
Area of Property:	Action Points/ Jobs/ Outcomes/ Tasks/ etc:

Additional Notes/ Initial Thoughts/ Potential Problems/ Follow Ups etc:

Clients Name:	
Clients Phone No. Email, skype etc.:	
Date of meeting:	
Brief Description:	
Project Name/ REF:	
Area of Property:	Action Points/ Jobs/ Outcomes/ Tasks/ etc:

Additional Notes/ Initial Thoughts/ Potential Problems/ Follow Ups etc:

Clients Name:	
Clients Phone No. Email, skype etc.:	
Date of meeting:	
Brief Description:	
Project Name/ REF:	
Area of Property:	Action Points/ Jobs/ Outcomes/ Tasks/ etc:

Additional Notes/ Initial Thoughts/ Potential Problems/ Follow Ups etc:

Clients Name:	
Clients Phone No. Email, skype etc.:	
Date of meeting:	
Brief Description:	
Project Name/ REF:	
Area of Property:	Action Points/ Jobs/ Outcomes/ Tasks/ etc:

Additional Notes/ Initial Thoughts/ Potential Problems/ Follow Ups etc:

Clients Name:	
Clients Phone No. Email, skype etc.:	
Date of meeting:	
Brief Description:	
Project Name/ REF:	
Area of Property:	Action Points/ Jobs/ Outcomes/ Tasks/ etc:

Additional Notes/ Initial Thoughts/ Potential Problems/ Follow Ups etc:

Clients Name:	
Clients Phone No. Email, skype etc.:	
Date of meeting:	
Brief Description:	
Project Name/ REF:	
Area of Property:	Action Points/ Jobs/ Outcomes/ Tasks/ etc:

Additional Notes/ Initial Thoughts/ Potential Problems/ Follow Ups etc:

Clients Name:	
Clients Phone No. Email, skype etc.:	
Date of meeting:	
Brief Description:	
Project Name/ REF:	
Area of Property:	Action Points/ Jobs/ Outcomes/ Tasks/ etc:

Additional Notes/ Initial Thoughts/ Potential Problems/ Follow Ups etc:

Clients Name:	
Clients Phone No. Email, skype etc.:	
Date of meeting:	
Brief Description:	
Project Name/ REF:	
Area of Property:	Action Points/ Jobs/ Outcomes/ Tasks/ etc:

Additional Notes/ Initial Thoughts/ Potential Problems/ Follow Ups etc:

Clients Name:	
Clients Phone No. Email, skype etc.:	
Date of meeting:	
Brief Description:	
Project Name/ REF:	
Area of Property:	Action Points/ Jobs/ Outcomes/ Tasks/ etc:

Additional Notes/ Initial Thoughts/ Potential Problems/ Follow Ups etc:

Clients Name:	
Clients Phone No. Email, skype etc.:	
Date of meeting:	
Brief Description:	
Project Name/ REF:	
Area of Property:	Action Points/ Jobs/ Outcomes/ Tasks/ etc:

Additional Notes/ Initial Thoughts/ Potential Problems/ Follow Ups etc:

Clients Name:	
Clients Phone No. Email, skype etc.:	
Date of meeting:	
Brief Description:	
Project Name/ REF:	
Area of Property:	Action Points/ Jobs/ Outcomes/ Tasks/ etc:

Additional Notes/ Initial Thoughts/ Potential Problems/ Follow Ups etc:

Made in the USA
Monee, IL
04 November 2026

67375797R00056